Reprints of Economic Classics

MATHEMATICAL INVESTIGATION

John Edward Tozer

MATHEMATICAL INVESTIGATION

OF THE

EFFECT OF MACHINERY

ON THE

WEALTH OF A COMMUNITY IN WHICH IT IS EMPLOYED

AND ON THE

FUND FOR THE PAYMENT OF WAGES

[1838]

ON THE EFFECT OF THE NON-RESIDENCE OF LANDLORDS ON THE WEALTH OF A COMMUNITY

[1840]

Two Papers from the *TRANSACTIONS* of the Cambridge Philosophical Society

Edited and With an Introduction by

DAVID A. COLLARD

REPRINTS OF ECONOMIC CLASSICS

Augustus M. Kelley · Publishers
NEW YORK 1968

First Published in *TRANSACTIONS* of the
Cambridge Philosophical Society
Volumes VI (1836-1838) & VII (1842)

Reprinted 1968 by
AUGUSTUS M. KELLEY · PUBLISHERS
New York New York 10010

Library of Congress Catalogue Card Number
66-21696

PRINTED IN THE UNITED STATES OF AMERICA
by SENTRY PRESS, NEW YORK, N. Y. 10019

INTRODUCTION TO THE WORK OF JOHN EDWARD TOZER

Tozer,[1] was a nineteenth century Fellow of Caius College, Cambridge. In 1838 and 1840 he produced two papers in an attempt to introduce mathematical analysis into political economy, at a time when Whewell of Trinity was trying to do the same thing. Unfortunately Tozer seems to have deserted economics, thereafter devoting himself to the law in which he achieved distinction. He also produced a rather eccentric paper in 1843 on the application of probability theory to the statements of witnesses and two (or perhaps three) papers, one on scientific method, towards the end of his life.

John Edward Tozer was born at Woolwich in 1806, the son of John Tozer (referred to simply as 'gentleman') and educated privately under a Mr. Henderson. At the age of twenty-six he was admitted to Caius, Cambridge, as a Pensioner. He did rather well at Mathematics, carrying off first prize in 1833 and 1834. In 1833 he was promoted to Scholar and graduated in 1836 as thirteenth "wrangler". It was in the four years after this that he wrote his two pieces on economic subjects, the papers on '*Machinery*' and '*Landlords*'.

The work of 1838 on '*Machinery*' is one of distinction, given its assumptions and the way in which the subject had previously been treated. The 1840 paper is also reproduced here for the sake of completeness but his other work is simply summarized in the Introduction.

The works that I know of are:—

1838 Mathematical Investigation of the Effect of Machinery on the Wealth of a Community in which it is employed, and on the Fund for the Payment of Wages.
(Read May 14, 1838—*Cambridge Philosophical Transactions*).

1840 On the Effect of the Non-Residence of Landlords, etc. on the Wealth of a Community.
(Read March 16, 1840—*Cambridge Philosophical Transactions*).

1843 On the Measure of the Force of Testimony in Cases of Legal Evidence. (Read November 27, 1843—*Cambridge Philosophical Transactions*).

1873 On Identity of Method in the Search for Truth-Jurisprudence (loose paper).

1874 On the Definition and Consolidation of the Law of Homicide (loose paper).

There may also be another paper titled 'On Probability' but I had not come across this at the time of going to press.

The little that I have gleaned about Tozer's life is drawn from a notice in the *Times* of December 13th, 1877, an obituary from the *Solicitors Journal* (Vol. 22) and the Gonville and Caius *Biographical History* (Vols. II and III).

I am grateful to the following for help at various stages:—Mr. Dobb of Trinity, Cambridge; Mr. Farrell of Gonville and Caius, Cambridge; the Librarians of Trinity and Gonville and Caius Libraries, Cambridge, Lincoln's Inn, London and the Cambridge Philosophical Society; the office of the Vice-Chancellor, Cambridge; the inter-library loan department of the University of Bristol.

The Classical Position on Machinery

The immense hardships suffered by the English working classes at the beginning of the nineteenth century were at least partially due to the introduction of labour-saving machinery. Rioting and machine-breaking were their only means of protest and popular history records the vindictive dispersion of demonstrators at St. George's Fields, Manchester, the March of the Blanketeers and Peterloo. The paradox of misery in a country of unprecedented wealth—a spectacle that distressed Sismondi among others—was most striking at the local level. Nearly a half of the population of Huddersfield, for instance, subsisted on twopence a day in 1829 at a time when output was at record levels.[(2)] Parliament was on the whole hostile to interference (especially suggestions of a tax on new machines) but there seems to have been strong feeling against the results of technical progress not only among the working classes but also among the smaller manufacturers. Certainly it is no tribute to the common sense of political economists of the stature of McCulloch and the younger Mill that they felt able to base their opposition to interference on *empirical* grounds. There was a clear conflict in this period between the requirements of economic growth and social justice and for all their "empirical" conjectures, political economists were implicitly making some fairly unpleasant value judgments about interpersonal and intertemporal utility.

Lord Lauderdale had taught in 1804 that capital displaced labour, referring to "the riots that have taken place on the introduction of machinery, and particularly at the time the ingenious machines for carding and spinning were first set a-going".[(3)] But the really important classical contribution, apart from Ricardo, came from John Barton whose tract on the "Condition of the Labouring Classes of Society" appeared in 1817. Barton started out from the assertion of the Poor Law reports that, as any relief must come from a fixed fund, there was no point in giving it and, further, that increased employment is directly linked to the general increase of wealth. Barton notices evidence from the reigns of Henry VII and post-Charles II as a refutation and argues that "as the doctrine that the progress of population is measured by the increase of wealth does not appear to be true in fact, so on the other hand it seems to me not consistent with sound reasoning". He then makes the crucial theoretical distinction that "the demand for labour depends on the increase of circulating, and not of fixed capital. . . . As arts are cultivated, and civilisation is extended fixed capital bears a larger and larger proportion to circulating capital". Thus, as Marx is to argue, the organic composition of capital increases through time. Perhaps even more important the ratio of fixed to circulating capital is determined by "the proportion which the wages of labour. . . . bear to the whole produce of that labour".(4) The *cause* of differing labour intensities did not, generally speaking, bother classical economists and can be left to one side. Thus Barton had by 1817 already presented the core of the intellectual case against machinery. In particular he had produced the kind of numerical example that was soon to be adopted by Ricardo.

Ricardo's attitude changed from sharp dogmatism in 1816 to mature and cautious pessimism in 1823. Rejecting Barton's analysis he claims in 1817 that 'on this subject I have completely satisfied myself' and in a parliamentary speech he declared that 'he never could think that machinery could do mischief to any country, *either in its immediate or permanent effect.*(5) But by 1823 this optimism had vanished. It still seemed to Ricardo that machinery was for the general good and should not be discouraged. Nevertheless it would cause a fall in the demand for labour and the only hope seemed to lie in workers' regulation of their own supply—"a little forethought, a little prudence—a little of that caution which the better educated felt it necessary to use".(6)

Both as a matter of his personal intellectual history and as the background to Tozer's own analysis, the transition is of the greatest interest. Mr. Sraffa has traced the first hint of Ricardo's changing attitude to one of Malthus's letters to Sismondi.(7) In the *Notes on Malthus* we find Ricardo taking over Barton's distinction; "to the capitalist it can be of no importance whether his capital consists of fixed or of circulating capital, but it is of the greatest interest to those who live by the wages of labour; they are greatly interested in increasing the gross revenue, as it is on the gross revenue that must depend the means of providing for the population".[8] The new doctrine appears in a fresh chapter "On Machinery" in the third edition of the *Principles.* Ricardo had formerly believed that whenever the *net produce* of society rose, the *gross produce* (wages-fund plus net produce) must also rise. He now abandons this belief and acknowledges that "the opinion entertained by the labouring class, that the employment of machinery is frequently detrimental to their interests, is not founded on prejudice and error, but is conformable to the correct principles of political economy".[9]

The injury, however, would only be short term. As for the longer run, Ricardo's reasoning is worth quoting in full. "As, however, the power of saving from revenue to add to capital, must depend on the efficiency of the net revenue, to satisfy the wants of the capitalist, it could not fail to follow from the reduction of the price of commodities consequent upon the introduction of machinery, that with the same wants he would have increased means of saving —increased facility of transferring revenue into capital. But with every increase of capital he would employ more labourers; and, therefore, a portion of the people thrown out of work in the first instance, would be subsequently employed; and if the increased production, in consequence of the employment of the machine was so great as to afford, in the shape of net produce, as great a quantity of food and necessaries as existed before in the form of gross produce, there would be the same ability to employ the whole population, and, therefore, there would not necessarily be any redundance of labour". That is, the use of machinery would increase the real incomes of capitalists and encourage saving and accumulation. And Ricardo seems to believe that alleviation of the labourers' lot would be sooner, rather than later. "These savings it must be remembered are annual, and must soon create a fund, much greater than the gross revenue, originally lost by the discovery of the machine, when the demand for labour will be as great as before and the situation of the people will be still further improved by the increased savings which the increased net revenue will still enable them to make". "These truths", he writes to McCulloch, "appear to me to be as demonstrable as any of the truths of geometry, and I am only astonished that I should so long have failed to see them".[10]

As one of the great propagandists of the Ricardian school, McCulloch's reaction was important. We must remember that he originally believed that as fixed capital displaces a larger volume of circulating capital it had the effect of lowering wages. Having been converted by Ricardo he now believed that "no improvement of machinery can possibly diminish the demand for labour"[11] and reacted with astonishment and dismay to Ricardo's new chapter. McCulloch begged Ricardo to reconsider the subject and in the subsequent exchange of letters only moved his position to the extent of admitting hardship for the labourers as a barely possible theoretical case.

With Ricardo dead, the orthodox 'Ricardian' view of 1825 is to be found in McCulloch's '*Principles*'. He admits that the capitalist would be indifferent between employing a fixed or a circulating capital as long as they gave the same profit. He grudgingly concedes that an innovation biased towards the use of fixed capital could cause unemployment. But there is no empirical evidence for this; it is 'extremely unlikely', 'barely possible'. An inveterate opponent of Malthus and Sismondi, McCulloch deploys as subsidiary arguments the labour needed to build machines and the increased demand for the now cheaper machine-made produce, but makes little of Ricardo's very important 'accumulation' argument. His conclusion is quite clear and contrasts sharply with Ricardo's doubts. "No introduction of machines having a tendency to lower the price and increase the supply of commodities, can possibly diminish the demand for labour, or reduce the rate of wages. The introduction of such machines into one employment, necessarily occasions an equal or greater demand for the disengaged labourers in some other employment. The only hardship which they ever impose of the labourer, is, that in some cases they force him to change his business".[12]

We must notice that in Mill's *Principles* (the first edition of which appeared in 1848) there was no sign of analytical advance in this part of the subject. After an ambivalent discussion Mill comes nearer to McCulloch than to Ricardo but is, in any case, able to take a rather lofty view because of his belief in the approaching stationary state. Roughly, he accepts the proposition that ill effects are theoretically possible but empirically unlikely. "All attempts to make out that the labouring classes as a collective body cannot suffer temporarily by the introduction of machinery . . . are, I conceive, necessarily fallacious . . . nevertheless . . . as things are actually transacted, improvements in production are not often, if ever, injurious, even temporarily, to the labouring classes in the aggregate".[13]

Marx, as is well known, makes the fixed-circulating distinction a central part of his system. An increasing organic composition of capital means that to absorb an expanding labour force, accumulation must take place at an ever increasing rate. As is not unusual in Marx, a rather dull looking technical distinction is successfully incorporated into a vigorous and exciting dynamics. In the neo-classical analysis, on the other hand, the whole subject was shorn of its glamour and turned into an exercise on factor substitutability and complementarity. Thus Marshall speaks of the "real and effective competition between labour in general and waiting in general. But it covers a small part of the whole field, and is of small importance relatively to the benefits labour derives from obtaining cheaply the aid of capital".[(14)] Wicksell asserts (referring to Ricardo) that 'the question has remained in this somewhat unsatisfactory position until the present time'. He has anxieties not about full employment (reintroduction of formerly 'inferior' techniques as wages fall will see to that) but about the behaviour of the marginal product of labour as technical progress takes place. But that is another story.[(15)]

Tozer on "Machinery"

The introduction of machinery involves a *flow* of investment, a *flow* of annual benefits, an increase in the *stock* of fixed capital and, most importantly, an initial fall and then an eventual rise in the wages-fund, itself a *stock* concept. Tozer's claim to a niche in the history of economic thought stands or falls with his treatment of this problem.

The following account, shorn of his algebra but illustrated by means of a simple diagram should make clear what Tozer was trying to do. His treatment of what we shall call the 'construction period' is not essential to the rest of the analysis but has one interesting feature. The total amount 'abstracted' from the wages fund is to include annual investments grossed up by the time they have been invested, the amount of 'grossing up' depending on the rate of profit. Labour will, of course, as McCulloch and others had pointed out, be used in the construction of machinery and the wages fund diminished by so much less. A formula* gives the total loss as depending upon:—

the total capital 'abstracted'

the rate of profit

the period of construction

the labour intensity of the machine industry.

---

*p. 510 line 6. In the first term q should replace x. See substitution from line 14 on previous page.

LOSS TO THE WAGES FUND

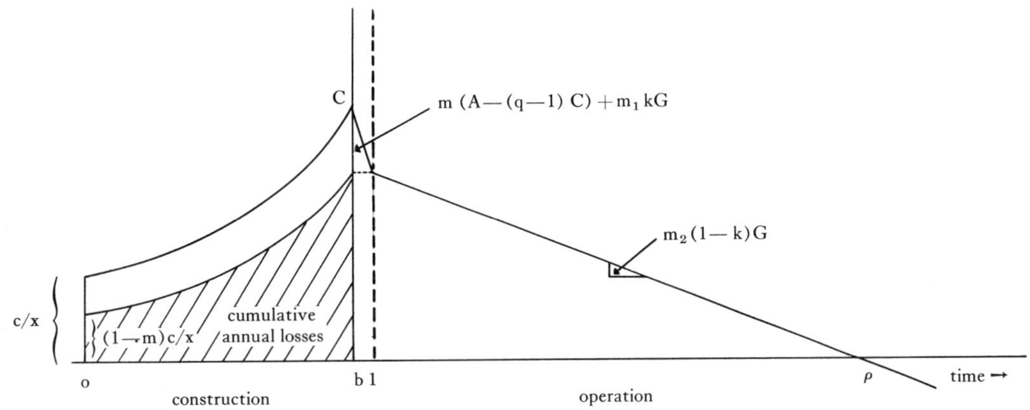

Referring now to the figure, the total loss to the wages fund during the construction period of b years is shown by the shaded area. The dotted curve shows the total capital 'abstracted' and rises to C.

We have spoken of the *flow* of annual benefits. Any gain can be looked upon either as an increase in productivity or as a fall in prices. In either case its measure is essentially the net value product of the machine; it depends upon the ratio between "the produce due to the use of the machinery" and "the produce due to the labour it displaces". Additionally, if costs are to be properly taken into account they too must be reduced to an "annuity". This is an annual sum sufficient to repay the ordinary profit on capital, replace the machinery and keep it in working order. Tozer's calculation of this annuity is a particularly sophisticated piece of work (p. 510) although he refers to it as the "common mode" (p. 514). The "whole annual gain to the community occasioned by" the machinery can now be written an annuity, G, which is the difference between two other annuities, one representing a stream of benefits and the other a stream of costs.

The annual gain, G depends upon:—

The total capital abstracted
the rate of profit
the productiveness of the machine
the production of the labour it displaces
the life of the machine
the annual expenditures for repairs
the scrap value of the machine

Having established the gain Tozer is now in a position to discuss the behaviour of the wages fund. The reader is again referred to the figure and to pages 511, 512. G "will be an annuity in money or other produce, of which a kth part may be consumed for immediate enjoyment and a (l-k)th part used as capital, also of the former an m_1th part*, and of the latter an m_2th part may go to the labourer". So, as in the earlier section on construction, there will be some employment of labour in the machine replacement industry. Now we come to the heart of the earlier controversy on the effects of machinery. There is an immediate loss to the wages fund of the whole capital "abstracted" (i.e., circulating capital converted into fixed). But this is lessened annually by $m_2(1\text{-}k)G$, that part of the annual gain (a flow) added to the wages fund (a stock). It is thus easily possible to derive a formula for ρ, the number of years before the wages fund is restored to its former level (ignoring the construction period). After this the fund *increases* by $m_2(1\text{-}k)$. Society as a whole will be interested in G but ρ is a rough indicator of the 'welfare' effects of introducing machinery.

The rest of the paper is concerned with minor modifications. We now go straight to the concluding paragraph which is quoted in full.

"The operation of the labourer is to abstract a fund which has been or would have been annually employed in the payment of wages, and annually renewed by the produce due to his exertions, and to supply a new fund, by increasing the wealth of the community, this portion is at first smaller than the fund abstracted, but it increases without any assignable limit, the rapidity of increase depending on the proportion in which the new fund is divided between the labourer and the other classes of society. Speaking with reference to the formulae, the rapidity of increase depends on the values of the arbitrary quantities, m, m_1, m_2, so these values can be assigned with greater or less exactness, as our statistical knowledge connected with the particular case is more or less accurate".

"Landlords"

The full title of Tozer's 1840 paper has an old-fashioned ring about it: "On the Effect of the Non-Residence of Landlords, etc. on the Wealth of a Community". Contemporary teaching was that it had little or no effect. Ricardo told the House (with respect to absentees living in Paris) that 'now, we had at any rate their capital, which was useful, though not so useful as if

*in the original this is mistakenly written as an mth part.

they also stayed at home'.[16] And Tozer, in passing, notices "the loss of the proprietor's services as a skilled labourer." But Mill, after a rather careful argument in one of his early essays, decided that, on balance, absentees neither harmed the home country nor benefited the foreign country.

We can profitably divide Tozer's treatment into two stages. The first is 'static' and largely reflects current doctrine. The second involves accumulation, or rather decumulation, and stems directly from the method employed in the paper on 'Machinery'. By definition, he points out, the landlord is simply the receiver of rent. If he moves away his rent will have to be exported to him. True to the Ricardian tradition, if exports are of raw material, rents rise and the rate of profit falls. Capital is diverted into less profitable uses and the community loses. If they are manufacturers, there is no effect on the rate of profit. "The accurate enunciation of the result appears to be, that beyond the loss to the revenue the absenteeism of proprietors can only impair the resources of a community when it forces capital from a more to a less profitable employment." Apart from minor embellishments concerning the revenues of the two governments, this summarises Tozer's contribution to the discussion at a static level.

But the discussion so far implies that the landlord expends "the whole income to which he is entitled, without leaving any trace of that expenditure in the shape of capital accumulated, with a view either to durable or prospective improvement, or to profit. This supposition must have been tacitly made whenever the wealth of a country has been pronounced to be unaffected by the non-residence of its landlords." If we now assume that the landlords would have used a proportion (la) of their incomes as capital and that this accumulation no longer takes place when they leave, then Tozer introduces a simple accumulation formula (similar to that on p. 509 of 'Machinery') to demonstrate how much, at the end of x years, the 'labour fund' will have diminished as a result of the landlord's absence. Alternatively, he works out the rather curious case where "the rate of accumulation [is] changed to a vth part of what it was in the preceding [year]". Finally, an allowance is made for any destruction of capital involved in the absence of landlords.

In all this there is no actual removal of capital from the country. It is simply that the rate of accumulation might be reduced, on the assumption that the landlord class had a relatively high propensity to save. "It is necessary to the truth of (the static) results, that the withdrawal of the proprietor should cause no removal of capital, that any part of the proprietor's income which was not expended should still be saved at home, and that no part should have been consumed without calling for the employment of capital. . . . If they be not fulfilled . . . the wealth of the community must be impaired by their absence; and the injury is capable of increasing with time to an indefinite extent".

'Landlords' is a lesser paper than 'Machinery' because although Tozer is very clear-headed about his analysis, the mathematical notation is not essential to the argument and is of doubtful expositional value.

Tozer's interests were quite clearly not confined to mathematics and political economy. In 1837 he was Mickleburgh Scholar in Chemistry but apart from familiarity with the subject displayed in the 1843 paper on 'Testimony' there is no further reference to any achievements in that line. He was appointed a Junior Fellow in 1838. An LL.D., he was called to the bar at Lincoln's Inn in 1840 becoming a member of the Norfolk Circuit and the Bedfordshire, Cambridgeshire, Huntingdonshire and Bury Sessions. He is said to have had considerable success as a junior. He remained, incidentally, unmarried.

"Testimony"

Three years after the paper on 'Landlords' Tozer made another mathematical incursion, this time into the legal field. Its subject matter was the measurement of the force of testimony in cases of legal evidence and, though by virtue of its subject matter it does not strictly concern us, it provides a further example of Tozer's optimistic faith in the fruitfulness of mathematical method.

The paper on 'Testimony' is a rather impractical exercise in applied probability theory. By this time a barrister-at-law, Tozer was anxious to demonstrate that one could, in principle, measure the force of testimony of witnesses by assigning numerical probabilities to their statements and then applying a general formula. He seems to be attacking two notions. The first is that one acquits where there is *any shadow* of doubt. "We may, perhaps, in criminal cases, make as small as we please the chance of an innocent person being convicted, but it must be either by increasing the chance of a guilty person escaping or by rendering the practical process more perfect."

The second is that probabilities are in any case inapplicable. On this he quotes Starkie: "the notion of those who have supposed that mere moral probabilities or relations could ever be represented by numbers or space, and thus be subjected to arithmetical analysis, cannot but be regarded as visionary and chimerical." Tozer disagrees: "if we were to conceive the elementary probabilities to be elicited by the skill of science, and presented to the jury as separate issues, they would then have to decide on simple facts instead of on a complete series of facts, and the remainder of the process would be logical deduction, and therefore would exclude the possibility of error. A result so obtained would possess all the accuracy of which the subject is capable: by as much as the practical process differs from this, by so much is it, as far as mere accuracy is concerned, inferior; the difference is the price of practicability."

A case of alleged arsenical poison is used for illustration. The probability that the wife is guilty is the probability that she knowingly possessed poison multiplied by the probability that, having the poison, she administered it. These probabilities are in turn built up from component probabilities that witnesses were telling the truth. As is made clear from the above quotation, the jury is to make up its mind on each component probability (with the aid of science where possible) and then apply Tozer's formula. If the probability of the wife being guilty turns out to be above some acceptable threshold value, she is convicted.

The present writer does not have sufficient qualifications either in law or in probability theory to give a ruling on the importance of this paper. On the face of it, it is a trivial exercise, the impracticability of which Tozer admits to in his introductory remarks; "practical men are concerned with practical rules, and with principles no further than may be sufficient to render those rules intelligible."

Further, his treatment of decisions in the face of uncertainty is rather cavalier.

Even so, the three papers together give a picture of one who, with missionary fervour, is attempting to bring mathematical reasoning into the very centre of the social sciences.

The Affair

In 1849 there took place a college dispute. I quote from his college's *Biographical History*.

"Two events occurred towards the close of his (Benedict Chapman, Master 1839-1852) life which seem to have greatly perturbed him. The first of these was the action of Mr. Tozer, a recently-elected senior fellow, who actually appealed to the Chancellor to exercise his authority in reforming certain points in the procedure of the college. It may be thought that this was no very serious offence, but the old master took it terribly in earnest. He wrote at once to the Chancellor, Prince Albert, pointing out that he was not visitor of the college, except in one special case. Mr. Tozer was then summoned before a special college meeting, solemnly reproved, and reminded that he had incurred the penalty of expulsion, and was only forgiven on his offering a humble apology. Technically the master was in the right*; but, considering in how many respects ordinary life in college departed from both the spirit and the letter of the ancient statutes, it seems almost grotesque to attach so much importance to a mistaken and unauthorized appeal to authority. How important the master thought this incident is shown by his recording it in full in the *Annals*, thus breaking the silence of two centuries; for this is the only entry in that volume since the time of the Commonwealth. This incident took place in 1849."

I have been permitted to see the *Annals*, in which a full account of the affair is given. Tozer seems to have been extremely determined at the outset but to have retreated in face of the unexpected zeal with which the old master defended his position. Through the medium of the Hon. Colonel Phipps, Tozer protested to Prince Albert that "I did not presume to address H.R.H. without having carefully ascertained that the Jurisdiction had been conferred on the Chancellor, and exercised by him without question of its legality, nor without having so determined the extent of the powers conferred on the authorities named by the Master, as to satisfy myself that those possessed by H.R.H. are exclusive." Dr. Chapman, now over 80 and deaf, consulted the records and statutes, producing evidence that apparently satisfied the Chancellor of the correctness of his position. Tozer, before being formally admonished by Master and Fellows, apologised, claiming that he had not intended to give offence but hinted darkly that he would make further moves if "compelled to do so in self-defence". Thus a rather silly episode was concluded. One may speculate that Whewell, as Master of Trinity, would not feel particularly favourably disposed towards Tozer.

The affair does not seem to have damaged his prospects within the University. He was appointed "Assessor" (legal advisor) to the Vice-Chancellor in 1849 and to the Chancellor in 1852. It seems that from about this period

* The Chancellor could be appealed to in the case of serious charges against the master, but not otherwise; and the appeal had even then to be made by a majority of the senior fellows.

Tozer became far more interested in the University than in the Law. "Mr. Tozer had formerly a good criminal business, but he had almost ceased to practice, being constantly resident at Cambridge, devoting his chief attention to his judicial duties as Chancellor's Assessor. He took a lively interest in all University questions and in scientific studies." We know that he abandoned circuit in 1858 having been appointed sergeant-at-law, "raised to the degree of the Coif."* From 1861 to his death he was Recorder (judge at Quarter Sessions) at Bury St. Edmunds.

Tozer and Whewell

It would not be extravagant to say that in his two papers on political economy Tozer was a disciple of William Whewell, polymath and Master of Trinity. Tozer certainly refers to Whewell's own papers in previous volumes of the *Cambridge Philosophical Transactions.* Both stressed the importance of the deductive method and both advocated a mathematical approach.

Whewell's work on political economy spanned the years 1829-1850[17]. His method is highly acceptable to the modern theorist. There are "two parts of the subject which it is of vital importance to keep separated;—the business of proving assumptions, and that of deducing their conclusions." Mathematics, of course, comes in with the second part. But his enthusiasm was sensibly qualified. Firstly he was prepared to accept the idea of equilibrium only as a first approximation (as in mechanics) to things, on the assumption that other effects are of an inferior order. Once granted "the postulate of equilibrium is introduced to our calculations by the process of putting the other postulates into the form of equations." Although he concedes that "a state of equilibrium is never attained, the state of equilibrium is that to which all the motions tend; and an equilibrium theory is the best attempt at a theory of trade which we can at present make." Secondly he has a fear of making the subject too mathematical. "Any attempt to make this subject at present a branch of Mathematics could only lead to a neglect or perversion of facts, and to a course of trifling speculations, barren distinctions, and useless logomachies."

Tozer was rather more optimistic about the use of mathematics (indeed the paper on 'Testimony' almost puts him on the lunatic fringe) provided that we neglect (a) "the happiness of mankind" and (b) the possibility of institutional reform modifying economic laws. This proviso is explicit in the paper on

*A Coif is the close-fitting cap of white lawn or silk worn by sergeants-at-law, barristers of the highest rank.

'Machinery' and is of the greatest importance. Nevertheless he sees economic science as a "series of propositions which are logical deductions from assumed definitions, and form those properties of the things defined that furnish axiomatic truths, and is therefore a subject to which mathematical reasoning is not only proper but necessary."

That they did not succeed in forming a flourishing school of mathematical economics is hardly surprising. As Schumpeter would say, they were 'men above their time'. In any case Whewell's energies were diffused over a vast area. In a single volume of the *Transactions* we find, apart from his 'Mathematical Exposition', three articles on various aspects of Plato, a criticism of Aristotle's account of induction and "The Transformation of Hypotheses in Science." Similarly Tozer was heavily engaged in legal duties, first in practice and then at the University. Neither was the sort of man who founds a school. Tozer had no influence whatever on the subject and lapsed into obscurity. Whewell is frequently mentioned but as a curiosity rather than as one of high stature. Marshall had certainly read later volumes of the *Transactions* (his signature appears in some of the volumes now at Bristol University) and it is probable that he benefited from Whewell's treatment of the role of the supply schedule and of the elasticity of demand. But this is not the place to examine Whewell's detailed contributions. Basically he gives an algebraic treatment to Ricardo and the younger Mill. But his introduction of the rising supply schedule as a general phenomenon, his careful period analysis of gold and price movements and his brilliant measures of demand elasticity are achievements of the highest order.

I find it strange that Whewell, in his 1850 papers, makes no reference to Tozer. Any suspicions that there might have been a coolness between them as a result of the 1849 incident is purely speculative. But there is no reference to Tozer in any of the catalogued Whewell papers at Trinity, Cambridge, nor have I discovered any such reference in a search of some of the uncatalogued papers. Certainly they formed no sort of alliance. We may conclude, fairly safely I think, that the embryo Cambridge School of mathematical economics died primarily at the hands of an indifferent and innumerate profession and less importantly because of the characters and scattered interests of its two protagonists.

The Final Years

Towards the end of a busy and practical life Tozer produced two or three other papers. The College biographical history seems to refer to two works, one

'On Probability' and the other *'On Identity of Method in the Search for Truth'*. The latter but not the former of these (if it is indeed a different paper from that of 1843) has come to light together with another paper *'On the Definition and Consolidation of the Law of Homicide.'* These papers appear to have been written in 1873 and 1874. Experience seems to have modified his earlier stress on the deductive method, particularly when backed by mathematical analysis. For instance in the paper on Homicide, Tozer argues that a very wide definition is best. To expand and modify the definition in order to include exceptions and peculiarities is to do in a formal way something that is much better done by the Courts as experience of individual cases accumulates. He uses the analogy of a geometer who tries to plot an undefined curve by getting as many points on it as possible rather than formulating an ever more complicated but, in the nature of things, never complete description.

Certainly this does not mean a reversion to crude induction. Rather, his is the respectable positivist view. He mentions (along with Jurisprudence which is by now his chief interest) Political Economy in which 'the balance and micrometer can be no longer used' and where 'we can neither make experiments at pleasure, nor watch the operation of nature's laws for the appearance of the phenomena whose happening we can with confidence predict'. It is a subject "in which the nearest approach to numerical accuracy is to be found in the columns of a statistical table." He asks the crucial question of whether the method of arriving at the truth in these (social) sciences is the same as in the natural sciences and answers that indeed it is. He speaks of applying the Baconian method to "Sciences that are not of numerical accuracy" and seizes on the need to test hypotheses in the face of experience. A "hypothesis cannot be treated as a fact, or as an established law, until it has been verified by additional experience—and it must be at once abandoned, if proved by such experience to be unsound" (later to be known as 'falsification'). Returning to his earlier concern with probability, Tozer claims that although one cannot predict about particular events "we can ascertain what the events are that tend to occur, and, with more or less accuracy, fix the limits within which that probability must lie." Thus he refers to a legal code as a "mere bundle of hypotheses." As observation modifies and takes the place of purely conjectural assumptions theories become more "plausible and comprehensive." Experience and observation not *'les ténèbres de l'antiquité'* are the sources of scientific progress.

Although the details of Tozer's paper on *Method* are a matter for the jurist rather than the economist, the contribution is of interest as a pioneering attempt to bring positivist scientific method into the social sciences just as in 1840 he had been trying to introduce mathematical reasoning. Whewell, of course, had already pioneered that explanation of scientific method now known as hypothetico-deductive.

John Tozer died on December 8th, 1877 at his residence, 4 St. James Street, Buckingham Gate.

DAVID A. COLLARD

University of Bristol

April 1968

NOTES

(1) Jevons refers to Tozer in his Mathematical bibliography (*Principles,* 3rd Edit.). This bibliography was reprinted in Fisher's edition of Cournot (1897). J. Viner also makes a passing reference to the paper on *'Landlords'* in his *Studies in the Theory of International Trade,* p. 321.

Since this introduction was prepared, my colleague Professor J. K. Whitaker has discovered some notes on Tozer's *'Machinery'* paper among the papers of Alfred Marshall at the Marshall Library, Cambridge. These notes consist of some quotations and a summary. Mr. Piero Sraffa has suggested to me, however, that the handwriting might be that of some person other than Marshall. At any rate the notes substantiate our conjecture that Marshall was familiar with the 'Transactions'.

(2) E. P. Thompson *The Making of the English Working Class,* Gollancz 1965. Chapter 9 gives a valuable account of the condition of the poor at this time.

(3) James Maitland, eighth Earl of Lauderdale *Inquiry into the Nature and Origin of Public Wealth* (1804).

(4) John Barton *Condition of the Labouring Classes of Society* (1817 and reprinted in 1934 with J. H. Hollander editing). After the early pages the work becomes relatively uninteresting from an analytical point of view.

(5) References to Mr. Piero Sraffa's edition of Ricardo are simply given as Sraffa. See Ricardo's letter to Barton, 20th May 1817 (Sraffa VII letter 218) and the press reports of his speech of 16th December 1819 on Robert Owen's plan for relieving the distress of the poor (Sraffa V p. 31).

(6) Speech on 30th May 1823 (Sraffa V, p. 303).

(7) Malthus to Sismondi 1821 (Sraffa VIII, pp. 375-7).

(8) Ricardo's *Notes on Malthus* (Sraffa II, pp. 350 et. seq.).

(9) See the celebrated chapter 31 of Ricardo's *Principles.*

(10) Ricardo to McCulloch 18th June 1821, (Sraffa VIII, p. 390).

(11) Edinburgh Review March 1821.

(12) McCulloch *Principles* 1825 (pp. 181-2).

(13) J. S. Mill *Principles* 1848. Ashley's 1909 Edition, pp. 93 *et. seq.* and pp. 742-3.

(14) A. Marshall *Principles* pp. 541-2.

(15) K. Wicksell *Value, Capital and Rent* and Lecture I especially pp. 135 *et. seq.* Wicksell challenged Ricardo's view that the 'gross' produce might fall.

(16) Speech on 16th May, 1822 (Sraffa V, pp. 186-8) on 'Absentees'. 600 inhabitants of Somerset were concerned about the tax-dodging activities of English families living in Paris.

(17) *Cambridge Philosophical Transactions* Vol. 3, "Mathematical Exposition of Some Doctrines of Political Economy"; Vol. 4, "Mathematical Exposition of some of the Leading Doctrines of Mr. Ricardo's 'Principles' "; Vol. 9, Second Memoir and Third Memoir. A reprinting of Whewell's work is forthcoming.

XXVI. *Mathematical Investigation of the Effect of Machinery on the Wealth of a Community in which it is employed, and on the Fund for the Payment of Wages.* *By* JOHN TOZER, ESQ. B.A. *of Caius College.*

[Read *May* 14, 1838.]

IN the third and fourth Volumes of the Society's Transactions are two papers by Mr Whewell, in which symbolical language is applied to the solution of some problems of Political Economy. In the following Paper another problem of the same science is subjected to a similar mode of investigation.

An opinion has been expressed that the term Political Economy has acquired an extent and a vagueness of meaning which in a high degree unfits it for the purposes of science. Certainly any attempt to apply mathematical reasoning to all the subjects which have by different writers, and at different times, been included in the name, must be altogether unsuccessful. Neither do we possess the data, nor has analysis the powers, necessary to such a task.

If, however, the investigation of the causes which affect the accumulation and distribution of wealth, be kept distinct from any considerations as to the effect of that accumulation, or the mode of its distribution on the happiness of mankind, and be also separated from any speculations or deductions as to the nature of those political and social institutions by which these causes may be modified or brought into action; the science that results, by whatever name it may be called, acquires an almost entirely demonstrative character—becomes a series of propositions which are logical deductions from assumed definitions, and form those properties of the things defined which furnish axiomatic truths, and is therefore a subject to which mathematical reasoning is not only proper but necessary.

The introduction of any process by which accuracy may be given to the reasonings of Political Economists, as tending, however remotely and indirectly, to place the science in this position, must be valuable; and even if we could suppose the principles to be as accurate in their enunciation, and as complete in their demonstrations as they can be rendered, there would still devolve on us the duty of rendering our deductions from them general, and of proving that they were necessary.

The particular problem under consideration, is of very limited extent, and of very easy solution. The method that has generally been employed has been to take particular numerical examples, and the results of these have frequently been assumed to lead legitimately to general conclusions. If the examples chosen had always been supplied by statistical facts, we should at least have been assured that the phenomena displayed in these results either had or might have happened, however unsatisfactory the general conclusions from them might have been.

This advantage, however, has not been afforded, the numbers have been generally assumed without reference to realities, and though it may sometimes have been carefully stated, that the conclusions could not possess a higher degree of truth than the premises, the impressions on the minds of general readers would be favourable to that particular conclusion which the example chosen tended to support.

PROP. A portion of capital, which either has been or would have been employed in the payment of wages, is used in the construction of machinery; to determine the effect on the wealth of the community, and on the fund for the payment of the labourer.

Let C be the capital,

$q-1$ the ordinary rate of profit.

If C be engaged in paying wages, it will yield annually a profit $= C(q-1)$; this, while the machinery is in progress, must be partly taken from capital.

Suppose the machinery to take b years to complete, and a part of $C = \frac{C}{x}$ to be expended every year in its construction till it is completed, then the productive capital will be diminished,

at the beginning of 1st year by $\frac{C}{x}$,

.... 2nd .... $\frac{C}{x}(1+q)$,

.... 3rd .... $\frac{C}{x}(1+q+q^2)$,

....

.... bth .... $\frac{C}{x}(1+q+\ldots\ldots q^{b-1}) = \frac{C}{x}\frac{q^b-1}{q-1}$,

.... $b+1$th...... $\frac{C}{x}\cdot q\frac{q^b-1}{q-1}$,

and at this time the whole capital will have been expended on the machine;

$$\therefore\ C = \frac{C}{x}q\cdot\frac{q^b-1}{q-1},\quad \text{and}\quad x = q\cdot\frac{q^b-1}{q-1}.$$

Let a fractional part, the mth of the expenditure for machinery, go to the labour, and B_1, $B_2 \ldots\ldots B_\varrho$ be the sums by which the fund for paying wages is diminished during the 1st, 2nd ...... ρth years of the progress of the machinery.

Thus

$$B_1 = \frac{C}{x}(1-m),\qquad B_2 = \frac{C}{x}(1+q-m);$$

$$B_3 = \frac{C}{x}(1+q+q^2-m),$$

$$B_\varrho = \frac{C}{x}\left(\frac{q^\varrho-1}{q-1} - m\right) = \frac{C}{q}\cdot\frac{q-1}{q^b-1}\left(\frac{q^\varrho-1}{q-1} - m\right).$$

The total loss to the labour-fund while the machinery is in progress is

$$\Sigma B = \frac{C}{x}\{1 + \frac{(q^2 - 1) + q^3 - 1 + \dots + q^b - 1}{q-1} - bm\}$$

$$= \frac{C}{x}\left\{\frac{q + q^2 + \dots + q^b - b}{q-1} - bm\right\}$$

$$= \frac{C}{x}\left\{q \cdot \frac{q^b - 1}{(q-1)^2} - b\left(\frac{1}{q-1} + m\right)\right\}$$

$$= \frac{C}{x}\cdot\frac{q-1}{q^b - 1}\left\{q \cdot \frac{q^b - 1}{(q-1)^2} - b\left(\frac{1}{q-1} + m\right)\right\}.$$

And in the 1st year of its employment the whole capital C is abstracted.

Let the machinery be made to last d_1 years by the expenditure of $a_1, a_2, a_3 \dots\dots a_{d_1}$ in the 1st, 2nd, d^1 years respectively and be worth a when rejected: also let the expenditure and profit be equivalent when the machine continues d years unimpaired, and then becomes useless.

This condition gives

$$a_1 q^{d_1} + a_2 q^{d_1 - 1} + \dots\dots + a_{d_1} q - a = C(q^{d_1} - q^d),$$

$$\text{or} \quad \Sigma(a_t q^{d_1 - t}) - a = C(q^{d_1} - q^d);$$

$$\text{and} \quad d = \log.\{(q^{d_1}) - \frac{1}{C}(\Sigma a_t q^{d_1 - t} - a)\}\frac{1}{\log. q}.$$

Let V_e be the exchangeable value of the machinery,

A the annuity necessary to pay the profit on C and provide new machinery at the end of d^1 years, paying also for the necessary repairs during this interval,

r_1, the produce due to the use of the machinery,

r .... to the labour it displaces,

p, p_1, prices of produce before and after its employment,

G the whole annual gain to the community occasioned by its uses,

$D_1, D_2, \ldots\ldots D_\rho$ sums by which labour-fund is diminished at the end of 1st, 2nd ...... ρth years after it comes into operation,

D diminution during the 1st year.

$$\text{Then}\quad V_\rho = C,$$

$$A = C.\frac{q-1}{1-q^{-d}}.$$

Also, since the gain to the community may be measured by the price that would have been paid for the produce r, minus the price that is actually paid for it when the profits of the capitalist have reached the average rate,

$$G = Cq.\frac{r_1}{r} - A = Cq\left(\frac{r_1}{r} - \frac{1-q^{-1}}{1-q^{-d}}\right).$$

Or, this gain may be measured by the saving in expenditure, added to the cost of the additional produce enjoyed, reckoning that cost at the original price;

$$\therefore\quad G = Cq - A + \frac{r_1 - r}{r}.Cq = Cq.\frac{r_1}{r} - A \text{ as before.}$$

This will be an annuity in money or other produce, of which a kth part may be consumed for immediate enjoyment and a $(1-k)$th part used as capital, also of the former an mth part, and of the latter an m_2th part may go to the labourer, whilst an mth part of the annuity A after deducting the profit $C(q-1)$ is employed in the same way. Before the use of the machinery the expenditure was C.

Hence, $D = C$,

$$D_1 = C - m\{A-(q-1)C\} - \{m_1 k + m_2(1-k)\}G,$$

$$D_2 = C - m\{A-(q-1)C\} - \{m_1 k + 2m_2(1-k)\}G,$$

....

....

$$D_\rho = C - m\{A-(q-1)C\} - \{m_1 k + \rho m_2(1-k)\}G$$

$$= C\left\{1 - m\frac{q-1}{q^d-1} - \{m_1 k + \rho m_2(1-k)\}\,q\left(\frac{r_1}{r} - \frac{1-q^{-1}}{1-q^{-d}}\right)\right\}.$$

When this becomes negative, the fund for the employment of labour will have become greater than it would have been if the machinery had not been constructed, that is, when

$$1 - m\frac{q-1}{q^d-1} - m_1 kq\left\{\frac{r_1}{r} - \frac{1-q^{-1}}{1-q^{-d}}\right\}$$

$$< \rho m_2(1-k)q\left\{\frac{r_1}{r} - \frac{1-q^{-1}}{1-q^{-d}}\right\},$$

$$\text{or } \rho > \frac{1 - m\,.\,\dfrac{q-1}{q^d-1}}{m_2(1-k)q\left\{\dfrac{r_1}{r} - \dfrac{1-q^{-1}}{1-q^{-d}}\right\}} - \frac{m_1}{m_2}\cdot\frac{k}{1-k}.$$

This advantage will increase by the addition of

$$C\,.\,m_2(1-k)\,q\left\{\frac{r_1}{r} - \frac{1-q^{-1}}{1-q^{-d}}\right\}$$

annually, and will continue to increase indefinitely.

We may in these formulæ, substitute for the produce in terms of the price.

We have $pr = Cq$, $p_1 r_1 = A = Cq\,.\,\dfrac{1-q^{-1}}{1-q^{-d}}$;

$$\therefore\ \frac{r_1}{r} = \frac{p}{p_1}\cdot\frac{1-q^{-1}}{1-q^{-d}};$$

$$\therefore\ G = Cq\left\{\frac{p}{p_1} - 1\right\}\frac{1-q^{-1}}{1-q^{-d}},$$

$$D\rho = C\left\{1 - m\frac{q-1}{q^d-1} - \{m_1 k + m_2(1-k)\rho\}\,q\,.\left(\frac{p}{p_1} - 1\right).\frac{1-q^{-1}}{1-q^{-d}}\right\}.$$

It may be observed that p_1 cannot be $> p$; if it were, more than the ordinary profit would arise from employing labour, and the machine would be superseded.

In general the motive of the capitalist in supplanting labour by machinery, is to procure for his capital more than the ordinary profit. If, then, he can raise his rate of profit to q_1, $A^1 = C\,.\,\dfrac{q^1-1}{1-q_1{}^{-d}}$ will be the annuity which the produce must realize. The above values of G, however, are correct whatever q may become, since the gains of the capitalist are included in those of the community at large.

In a State where there is any number of machines at work, and also any number in progress, wealth will be increasing more rapidly than it would have been if all the capital had been employed in paying wages if ΣG be positive, and less rapidly if ΣG be negative, that is, more or less rapidly as,

$$\Sigma\frac{r_1}{r}.C > \text{ or } < \Sigma\frac{1-q^{-1}}{1-q^{-d}}.C, \text{ or as } \Sigma\frac{p.C}{p_1(1-q^{-d})} > \text{ or } < \Sigma\frac{C}{1-q^{-d}}.$$

If in any case the machinery lower prices, the community must gain by its use, for if $p_1 < p$, G is positive.

This supposes p_1 to give the capitalist no additional profit, *à fortiori*, therefore the community will gain, if, when the rate of profit is raised to $q' - 1$, p_1 be $< p$.

Let p'' be the price that would pay ordinary profits $q - 1$.

$$\text{Then } p''r_1 = C.\frac{q-1}{1-q^{-d}}; \quad p_1r_1 = C.\frac{q_1-1}{1-q_1^{-d}},$$

$$p'' = p_1.\frac{q-1}{q^1-1}.\frac{1-q_1^{-d}}{1-q^{-d}}.$$

The community will gain by the use of machinery, if

$$p > p_1\frac{q-1}{q^1-1}.\frac{1-q_1^{-d}}{1-q^{-d}}.$$

If a portion of the dispossessed labourers, whose wages were paid by a part $l.C$ of the capital, become dependent on society for support, there will result the equation

$$G + lC = Cq\left\{\frac{r_1}{r} - \frac{1-q^{-1}}{1-q^{-d}}\right\},$$

$$G = C\left\{q\frac{r_1}{r} - \left(\frac{q-1}{1-q^{-d}} + l\right)\right\}.$$

And the community would gain or lose as

$$\frac{r_1}{r} > \text{ or } < \frac{1-q^{-1}}{1-q^{-d}} + \frac{1}{q} l.$$

The common mode of calculating the annuity A has been used, but it does not appear to be strictly applicable; it makes no provision for paying machine makers in the first year after the machinery comes into use, and does provide for their payment in the $(d+1)$ year; that is, after the machinery is finished.

To commence the manufacture of new machinery immediately an additional capital must be employed, to advance the wages of its makers; and if the same sum is paid every year in this way, the additional capital must equal that part of the annuity A which is available for a similar purpose.

Let Cy be the additional capital.

Cy is the annuity, commencing at the beginning of 1st year and ending at the beginning of dth year, which pays for the machine.

Hence, since at the beginning of dth year this amounts to

$$Cy\,(1+q+\dots\, q^{d-1}) = Cy \,.\, \frac{q^d-1}{q-1},$$

and at the end of dth year this is worth $Cy\,.\,q\,.\,\dfrac{q^d-1}{q-1}$, which must $= C$,

$$y = \frac{1-q^{-1}}{q^d-1}.$$

Again, the produce must supply an annuity A, which pays the profits on $C(1+y)$, and accumulates by the end of dth year to $C(1+y)$.

The accumulation is therefore worth

at end of 1st year $A - C(1+y)(q-1)$,

.... 2nd year $\{A - C(1+y)(q-1)\}(1+q)$,

....

.... d^{th} year $\{A - C(1+y)(q-1)\}\dfrac{q^d-1}{q-1}$,

which must $= C(1+y)$.

This gives

$$A = \frac{q-1}{q^d-1} . C(1+y)\, q^d = C . \frac{1-q^{-1}}{1-q^{-d}} . \frac{q^{d+1}-1}{q^d-1},$$

$$G = Cq(1+y)\frac{r_1}{r} - A = Cq(1+y)\left\{\frac{r^1}{r} - \frac{1-q^{-1}}{1-q^{-d}}\right\}$$

$$= Cq . \frac{1-q^{-(d+1)}}{1-q^{-d}}\left\{\frac{r_1}{r} - \frac{1-q^{-1}}{1-q^{-d}}\right\}.$$

The whole capital employed is

$$C(1+y) = C . \frac{1-q^{-(d+1)}}{1-q^{-d}},$$

$$D\rho = C\frac{1-q^{-(d+1)}}{1-q^{-d}}\left\{1-m\frac{q-1}{q^d-1} - \{m_1k + m_2(1-k)\rho\}\, q\left(\frac{r_1}{r} - \frac{1-q^{-1}}{1-q^{-d}}\right)\right\},$$

$$D = C(1+y) - Cy = C.$$

Also

$$p_1r_1 = A = C(1+y) . \frac{q-1}{1-q^{-d}}; \quad pr = Cq(1+y);$$

$$\therefore \frac{r_1}{r} = \frac{p}{p_1} . \frac{1-q^{-1}}{1-q^{-d}};$$

and therefore,

$$G = Cy . \frac{1-q^{-(d+1)}}{1-q^{-d}} . \frac{1-q^{-1}}{1-q^{-d}}\left\{\frac{p}{p_1} - 1\right\},$$

and $D\rho = C\dfrac{1-q^{-(d+1)}}{1-q^{-d}}\left\{1-m . \dfrac{q-1}{q^d-1} - \{m_1k + m_2(1-k)\rho\}\left(\dfrac{r^1}{r} - \dfrac{1-q^{-1}}{1-q^{-d}}\right)\right\}.$

As before, the condition necessary that the community shall gain by the use of machinery is

$$p > p_1 . \frac{q-1}{q_1-1} . \frac{1-q_1^{-d}}{1-q^{-d}},$$

and since p^1 cannot be $> p$, the community cannot lose, unless q_1 be $< q$, or the capitalist also lose.

From the result of the above investigation it would seem that we are entitled to draw the following conclusions:

If we assume that a capitalist will employ machinery or labour as the one or the other will procure for him the highest rate of profit, then the employment of machinery will always increase the wealth of the community. Not only is the capitalist unable to secure his own advantage at the expense of any other class, he cannot even prevent a general participation in the benefit.

The operation on the labourer is to abstract a fund which has been or would have been annually employed in the payment of wages, and annually renewed by the produce due to his exertions, and to supply a new fund, by increasing the wealth of the community, a portion of which will in general be paid as wages; this portion is at first smaller than the fund abstracted, but it increases without any assignable limit, the rapidity of increase depending on the proportion in which the new fund is divided between the labourer and the other classes of society. Speaking with reference to the formulæ, the rapidity of increase depends on the values of the arbitrary quantities m, m_1, m_2, k, and these values can be assigned with greater or less exactness, as our statistical knowledge connected with the particular case is more or less accurate.

APPENDIX.

EXAMPLES extracted from writers on Political Economy.

M^c CULLOCH, 2d Edition, 193.

Comparison of two machines, one of which lasts one year the other ten years, the cost of each being £.20000, and their produce the same, the ratio of profit 10 per cent.

When $d = 1$,

$$y = \frac{1-q^{-1}}{q-1} = q^{-1} = \frac{10}{11}, \quad A = C(q+1) = C\frac{21}{10} = 42000.$$

When $d = 10$, let y and A become y^1 and A_1,

$$y^1 = \frac{1-q^{-1}}{q^{10}-1} = \frac{1}{11} \cdot \frac{10^{10}}{11^{10}-10^{10}} = \frac{5}{88} \text{ nearly},$$

$$A^1 = C \cdot \frac{1-q^{-1}}{1-q^{-10}} \cdot \frac{q^{11}-1}{q^{10}-1} = \frac{155}{844} C \text{ nearly} = 3672.$$

If G now represent the gain to the community by employing the more durable machine,

$$G = A - A_1 = \frac{8087}{4220} \cdot C = 38327.$$

Again, when the machine lasted 1 year there was paid in wages $mCy = \frac{10}{11} mC$. When 10 years, the amount paid in the ρth year of its being employed will be

$$mCy^1 + \{m_1 k + m_2 (1-k)\rho\} G = C\left\{\frac{5}{88} m + \{m_1 k + m_2 (1-k)\rho\} \frac{8087}{4220}\right\}.$$

Therefore if D_ϱ be the loss to the labour-fund at the end of the ρth year by employing the more durable machine,

$$D_\varrho = C\left\{m\left(\frac{10}{11} - \frac{5}{88}\right) - \frac{8087}{4270}\{m_1 k + m_2(1-k)\rho\},\right.$$

$$D_0 = Cm\frac{75}{88}.$$

If $m_1 = m_2 = m = k = \frac{1}{2}$; D_1 will be negative, and the labour-fund will be increased at the end of the 1st year.

SISMONDI.

An improvement takes place in machinery which reduces prices 5 per cent.

Suppose the produce the same.

$$\text{The condition gives } \frac{Cq}{A} = \frac{100}{95} = \frac{20}{19}, \quad A = Cq.\frac{19}{20},$$

$$G = Cq - A = Cq.\frac{1}{20},$$

$$D_\varrho = C - m\{A - C(q-1)\} - G\{m_1 k + m_2(1-k)\}\rho$$

$$= C\left\{1 - m\left(\frac{19}{20}q - q + 1\right) - \frac{q}{20}[m_1 k + m_2(1-k)\rho]\right\}.$$

If $m = m_1 = m_2 = k = \frac{1}{2}$, and $q = \dfrac{11}{10}$.

$$D_\varrho = C\left\{1 - \tfrac{1}{2}.\frac{200-11}{200} - \frac{11}{200}.\frac{1}{4}(1+\rho)\right\}$$

$$= \frac{C}{800}\{411 - 11\rho\}.$$

If $\rho > \dfrac{411}{11}$ the labour-fund will have gained.

Hence, after 38 years the amount spent in wages is greater than it would have been if the improvement had not taken place, and increases beyond this period to an indefinite amount.

The values assigned to m_1 &c. are entirely arbitrary, if m, m_1, m_2, each $= \frac{5}{6}$, and $k = \frac{1}{2}$,

$$D\rho = \frac{C}{480}(91 - 11\rho);$$

and the labour-fund will be increased after 9 years.

In the value of $D\rho$ the whole of C is supposed to be taken from the labour-fund, the data necessary for calculating the effect of the change enunciated in the proposition are not given.

RICARDO, 3d Edition, 469.

A capitalist employs £.20000, £.7000 of which is vested in irremoveable capital, and £.13000 in labour, profits being 10 per cent. He then abstracts £.6500 to pay machine makers, and £.1000 for half the profits during the year the machinery is in progress. His capital is now £.6500 to pay wages, and £.7500 in machinery, in addition to the £.7000 irremoveable.

The machinery is supposed indestructible, and no more is produced by it than will pay the common profits on the £.7500.

Take the capital as £.13000, and suppose that with the aid of the £.7000 profits are raised on this from $q-1$ to $q_1 - 1$.

$$\text{Then } q_1 - 1 = \frac{20000}{13000}(q-1) = \frac{20}{13}(q-1).$$

Now before the machinery was employed the produce of the £.7500 must have fetched $7500 . q$, and afterwards $7500 . (q_1 - 1)$;

$$\therefore \frac{r_1}{r} = \frac{q_1 - 1}{q_1} = 1 - q_1^{-1},$$

$$\text{and } G = Cq_1\left\{\frac{r_1}{r} - \frac{1-q_1^{-1}}{1-q_1^{-d}}\right\} = Cq_1\{1 - q_1^{-1} - (1 - q_1^{-1})\} = 0.$$

Since $d = D$,

$$D\rho = C\left(1 - m\frac{q-1}{q^d - 1}\right) - \{m_1 k + m_2(1-k)\rho\}\, G = C.$$

If therefore it were possible that the capitalist could be influenced by a motive which would induce him to employ machinery under such circumstances as these, the community would gain nothing, and the labourer would lose the whole.

BARTON.

A manufacturer has £.1000, with which he pays 20 men £.50 each; his capital is suddenly increased to £.2000, when he lays out £.1500 in machinery, which takes 1 year to complete, and does the work of 15 men, but requires 1 man to keep it in repair, and also lasts 15 years, profits 10 per cent.

Here $C = 1500$, $b = 1$, $d = 15$, $q = \frac{11}{10}$;

$$\therefore\ x = q\frac{q^b - 1}{q - 1} = q = \frac{11}{10},$$

$$B = (1-m)\frac{C}{x} = (1-m)\frac{15000}{11}.$$

But £.500 only of the £.1500 was before employed in labour, and therefore the loss to the labour-fund during the year that the machine is making, is

$$(1-m)\frac{15000}{11} - 1000 = \frac{1000}{11}(4 - 15m),$$

which is negative if m be $> \frac{4}{15}$, or if the labourer receive more than $\frac{4}{15}$ of the price of the machine; the amount paid for wages will be increased.

After the completion, we have

$$A = 1500 . \frac{\frac{11}{10} - 1}{1 - \left(\frac{10}{11}\right)^{15}} = 150 . \frac{11^{15}}{11^{15} - 10^{15}} = 197,$$

and the machine does the work of 14 men, for whose labour the consumer paid $50 . 14 . \frac{11}{10} = 770$;

$$\text{the gain in price is } 770-197=573,$$

$$\text{and the capitalist gains } 100;$$

$$\therefore\ G=100+573=673.$$

There has been taken from the labour-fund £.500.

There is returned to it $m(197-150)+\{m_1k+m_2(1-k)\rho\}\,673$.

Whence $D_\rho=500-(47m+673m_1k)-673m_2(1-k)\rho$.

$$\text{If } m=m_1=m_2=k=\tfrac{1}{2},$$

$$D_\rho=\frac{1233-673\rho}{4}.$$

The labour-fund gains after the 2nd year, and receives an addition of $£\,\dfrac{673}{4}$ every subsequent year.

In this solution the value of A, taken by Mr M^c^Culloch (not B) in refuting Mr Barton's views, is employed.

Taking what I have assumed to be a more correct value, we get,

$$\text{since } y=\frac{1-q^{-1}}{q^d-1}=\frac{9}{75}\text{ nearly,}$$

$$\text{and } A=C.\frac{q-1}{1-q^{-d}}(1+y)=220.$$

The capital employed in the machinery $=C(1+y)=1680$; and therefore it must produce as much as £.680 paid in wages.

$$\text{Hence, gain in cost of produce}=\frac{11}{10}.\,680-220=528,$$

and the capitalist gains 100;

$$\therefore\ G=528+100=628.$$

Again, there has been taken from the labour-fund,

$$C(1+y)-1000=1500.\frac{84}{75}-1000=£.680;$$

there is returned to it

$$mCy+\{m_1k+m_2(1-k)\rho\}\,G=m180+\{m_1k+m_2(1-k)\rho\}\,628$$
$$=4\,[157m+\{m_1k+m_2(1-k)\rho\}\,157];$$

$$\therefore\ D_\rho = 4\left[170 - 45m - \{m_1 k + m_2(1-k)\rho\}157\right],$$

$$D_0 = 4(170 - 45m).$$

If $m = m_1 = m_2$,

$$D_\rho = 4\{170 - m(45 + 157k) - m(1-k)\rho 157\};$$

and if each of these and also $k = \frac{1}{2}$,

$$D_\rho = 4(170 - \frac{247}{4} - \frac{157}{4}\rho) = 333 - 157\rho.$$

The labour-fund will have gained at the end of the 3rd year.

The produce is here supposed the same before and after the machine was employed. The machine must therefore do the work of as many men as the capitalist is deprived of the means of paying.

XI. *On the Effect of the Non-Residence of Landlords, &c. on the Wealth of a Community.* *By* J. TOZER, ESQ. M.A. *Caius College.*

[Read *March* 16, 1840.]

THE investigations that have been made by political economists of the effects produced on the wealth of a community by the non-residence of its proprietors, have frequently been asserted to furnish results which are not confirmed by observation. The following is believed to be a more careful investigation of the problem than any that has yet been made, and one that accounts for the apparent discrepancy.

While the proprietor resides, his income, subject to such deductions as are made by direct taxation, will be expended either in purchasing commodities or in paying for services. Those whose services he retains will expend what they receive in the same manner, and therefore the whole income of the proprietor will be expended, either directly or indirectly, in the purchase of commodities. The necessary and sufficient division of these will be into two classes, those which have been produced by the labour and capital of the countrymen of the proprietor, and those which have been produced by the labour and capital of foreigners, and which have been placed within his reach by the employment of capital and labour which may have been either native or foreign.

Of taxes, we need only consider those which the proprietor is constrained to pay while he is resident, and whose payment he evades by non-residence. We may also, without affecting the result, assume these to be paid when the income is realized, and not before.

We have then, while the proprietor resides, a portion of capital employed in raising the produce of his estates; a second portion in raising such other native commodities as are consumed by himself or those whose services he retains, and in distributing such foreign productions as are so consumed; and a third portion, which is commercial capital, belonging either to resident natives or to foreigners, and which is employed in purchasing and importing those foreign productions.

Each of these classes is susceptible of the usual further divisions into fixed and floating.

Call these portions of capital $C_1 C_1'$, $C_2 C_2'$, $C_3 C_3'$ respectively; $C_1' C_2' C_3'$ being fixed, and $C_1 C_2 C_3$ floating.

Then if q be the amount of a unit of capital with its profit, C_1, C_2, C_3, must at the end of the year, or other period of return, yield qC_1, qC_2, qC_3, respectively, and C_1', C_2', C_3' must yield annuities which pay the profits $(q-1)C_1'$, $(q-1)C_2'$, $(q-1)C_3'$, and replace such portions of those capitals as have been destroyed. Call these annuities A_1, A_2, A_3, respectively.

$$\text{Then if } q \,.\, C_1 + A_1 = q(c_1 + c_2 + \dots c_n) = q\Sigma c,$$

where $c_1, c_2 \dots$ are portions of capital which respectively yield the returns $r_1, r_2 \dots r_n$ for each unit, the fixed capital being expressed in terms of its value as floating capital, and $r_1, r_2 \dots r_n$ being a decreasing series, the proprietor's income will be

$$\Sigma_n^1 c(r-q), \text{ which suppose } = R;$$

Σ_n^1 indicating the sum of the terms $c_1(r_1-q)\; c_2(r_2-q) \dots c_n(r_n-q)$.

Of this, a part Rt suppose will be paid in those direct taxes the payment of which will be evaded by non-residence.

A part $qC_2 + A_2$ will be received by the owners of $C_2 + C_2'$.

The remainder $q \,.\, C_3 + A_3$ by the owners of $C_3 + C_3'$. Let this remainder $= \Sigma\rho$; ρ_1, ρ_2, &c. being portions of it which are exchanged for portions of foreign produce of different kinds.

Suppose now H to be the country in which the proprietor's lands are situated, K that in which the equivalent for ρ_1 is produced; and let σ_1 be the price of that equivalent in K.

Then if there be direct intercourse between H and K, and if K import the produce of the proprietor's estates, this price in H must have been raised to $\sigma(1+\epsilon+\eta)$ where $\epsilon\sigma$ is the expence of making the transfer, and $\eta\sigma$ the import duty in H, it being supposed that there are no export duties. If K do not import the produce of the proprietor's estates, she will either import some other articles produced in H, or the produce of some other country which does import from H; and there will be a series of exchanges effected by equivalents that are produced by capital other than that we are considering, the last operation of importing to H yielding a tax $\tau\sigma$ to its revenue.

Hence while the proprietor remains at home the produce of his estate is thus distributed:—

To native capitalists and labourers, $\Sigma_n^1 c\{r(1-t)+qt\} - \Sigma\rho$, of which $C_1 + C_2 + C'_1 + C'_2 + A_1 + A_2 - (C'_1 + C'_2).q$ is employed in replacing capital.

To revenues of foreign states, $\Sigma\sigma\Sigma\eta$.

To revenue of H, $t\Sigma_n^1 c(r-q) + \tau\Sigma\sigma$.

To commercial capitalists and labourers, $\Sigma\sigma\Sigma\epsilon$; the remainder of $\Sigma\rho$ replacing the commercial capital with which the foreign purchases were made.

When the proprietor becomes non-resident the capital $C_2 + C'_2$ will be disengaged, because his absence destroys the demand on which its employment depended; but a new demand for such commodities as can be exported with advantage will be created by the absence, because the rent of the proprietor must now be exported. It is therefore in raising such commodities that the disengaged capital will be employed. If then the exports of H be made in manufactured goods, there will not in general be any variation in the rate of profit, because the employment of additional capital in manufacturing does not diminish profits, if the demand for manufactures be proportionably increased.

The income to support the inhabitants of H is therefore precisely the same before as after the removal; but there is this difference, the equivalent for the produce of $C_2 + C'_2$ was then possessed by the proprietor, it has now to be created by the labour of those who produce the commodities, by the exportation of which his rent is paid.

Again, let the exports of H consist of raw produce, then to the series $c_1\, c_2 \ldots c_n$ there will be in general added the terms $c_{n+1}\, c_{n+2} \cdots c_{n+m}$, producing respectively the returns $r_{n+1} \ldots r_{n+m}$, an effect which will not be confined to the estate of the proprietor who has become an absentee. The rate of profit will be lowered from $q - 1$ to $q' - 1$, where q' is determined by the least return which is now made by an unit of capital; the whole rental of H will be increased, and the income of the proprietor will be raised from

$$R = \Sigma_n^1 c(r - q), \text{ to } R_1 = \Sigma_{n+m}^1 c(r - q')$$

$$\text{whence } R' - R = \Sigma_{n+m}^{n+1} c(r - q') + (q - q')\Sigma_n^1 c,$$

which will be the gain of the proprietor. The treasury of H will lose $t\Sigma_n^1 c(r - q) + \tau\Sigma\sigma$. Foreign treasuries will gain $R'\Sigma\eta' - \Sigma\rho\Sigma\eta$, $R'\eta_1'$, &c., being the import duties successively paid on R' and its equivalents.

If the same amount of commercial capital be required to export R' as was required to export $\Sigma\rho$, and import its equivalent, $C_3 + C''_3$ will be unaffected; but if a different amount, and if the owners of this capital be residents of H, then a portion of $C_3 + C''_3$ may be diverted to the same employments as $C_2 + C''_2$ have been compelled to seek; or a portion of $C_2 + C''_2$ may be employed as commercial capital. The effects of non-residence will thus be increased or diminished in degree, but will continue to be the same in kind.

The distinction between a country which exports manufactures and one that exports raw produce is not a necessary one, though it may generally exist. The accurate enunciation of the result appears to be, that beyond the loss to the revenue the absenteeism of proprietors can only impair the resources of a community when it forces capital from a more to a less profitable employment.

As regards the effect on F, the country to which the proprietor has removed, the nature of the products in which his rent will be imported is independent of the nature of those he consumes, as well as of those in which it was produced. The presence of an individual who is without capital, but who is entitled to an income, will therefore create a demand for the employment of capital in producing the commodities he requires.

This capital will be drawn from investments where its employment produces the least return. There will be a gain in direct taxes paid by the immigrant, and beyond this there will be a gain, if supplying his wants afford a more profitable employment for capital than that from which it is withdrawn. This will not necessarily be the case, since the importation of the income may have rendered some species of home production unprofitable, and the demand for capital being thus accompanied by its disengagement, the rate of profit may be unaffected. In the case in which F imports raw produce, there will be an abstraction of capital from its production at home.

If then $c_1 c_2 \ldots c_\nu$ be portions of capital, the units of which yielded before the withdrawal $r_1 r_2 \ldots r_\nu$ respectively of produce, $r_1 r_2 \ldots r_\nu$ being a decreasing series, and if $c_{\mu+1} \ldots c_\nu$ be withdrawn, the rate of profit being changed from q to q_1, the whole rental of F will be reduced from

$$\Sigma_\nu^1 c(r-q), \text{ to } \Sigma_\mu^1 c(r-q_1),$$

and therefore diminished by

$$\Sigma_\nu^{\mu+1} c(r-q) + (q'-q)\Sigma_\mu^1 c.$$

If the imports of F be made in manufactures, there will not in general be any increase of wealth consequent on the residence in F of the proprietor whose home is H, beyond the direct taxes he pays; but there will be a substitution of a portion of income obtained without labour for an equal portion obtained by labour.

The reason why these results contradict general opinion on the subject is this: we are compelled, as preliminary to our investigations, to limit by definition the meaning of the terms we employ. The proprietor, in the vocabulary of the political economist, is simply the individual entitled to a certain income when it can be realized, and constantly either anticipating that income, or devoting it when received to the purchase of products whose creation rendered the employment of the capital of others necessary; his absence therefore does not involve any removal of capital, and consequently does not diminish the means of supporting human life. It has also in this investigation been supposed that he and his dependents expend the whole income to which he is entitled, without leaving any trace of that expenditure in the shape of capital accumulated, with a view either to durable or prospective improvement, or to profit. This supposition must have been tacitly made whenever the wealth of a country has been pronounced to be unaffected by the non-residence of its landlords.

Let us now suppose a portion of the proprietor's income to have been, by himself or by his dependents, either accumulated or itself employed as capital in producing the commodities for which it is exchanged. Call this portion a, and let a part of it $(1-l)\,a$ be expended without accumulation, and a part la be employed as capital with a view to profit. If then the rate of profit continued constant, and the whole of the proceeds of la were employed as capital, we should have had in the $x+1^{\text{th}}$ year to expend on labour, instead of a,

$$a\left\{l\frac{q^x-1}{q-1}+(1-l)\right\};$$

and the fund for employing labour will therefore have been at the commencement of this year diminished by $qla\,.\,\dfrac{q^{x-1}-1}{q-1}$.

Or, if we suppose the rate of accumulation in any one year to be changed to a vth part of what it was in the preceding, from an alteration in the rate of profit, or from a different proportional part of that which is produced being accumulated; the fund for employing labour will be at the end of the first, second,.... years, instead of

$a,\ a\{1+lq\},\ a\{1+l(q+q^2)\}$, &c. $a,\ a(1+lqv),\ a\{1+l(qv^2+q^2v^{1+2})\}$, &c.;

and the loss to this fund during the $x+1^{\text{th}}$ year would be

$$a\{1+l\Sigma(qv^{\frac{x+y-1}{2}})^{x-y}\},\ y \text{ to be taken from } x-1 \text{ to } 1.$$

This expression may be made to include the loss of the proprietor's services as a skilled labourer.

But further, a consequence of the absence of the proprietor may be the removal of a portion of the capital which that absence has forced into new investments, and the destruction of another portion. Of the capital C_2+C_2', then, let the part mC_2+nC_2' be removed without a change of residence of the owners, and the part $m'C_2+n'C_2'$ be either destroyed or removed in such a way, that its profit is no longer expended in H; and let the fractional parts λ, λ' respectively of these have been employed in producing capital, the remainder having been expended without accumulation.

Then the resultant loss to the community from the latter part will be

$$(1 - \lambda) \{m C_2 + n [A_2 - (q - 1) C_2']\} + (1 - \lambda') (m' C_2 q + n'. A_2);$$

and from the former part, at beginning of 1st year,

$$\lambda m C_2 + \lambda' m' C_2;$$

at beginning of 2nd year,

$$\lambda m C_2 . q + \lambda' m' C_2 . q + \lambda n \{A_2 - (q - 1) C_2'\} + \lambda' n_1 A_2;$$

at beginning of 3rd year,

$$\lambda m C_2 q^2 v + \lambda' m' C_2 q^2 v + \lambda n \{A_2 q - (qv - 1) C_2' q\} + \lambda' n_1 A_2 qv;$$

....

at beginning of $x + 1^{th}$ year,

$$\{(\lambda m + \lambda' m') C_2 q v^{x-1} + (\lambda n + \lambda' n' v^{x-1}) A_2 - \lambda n (q v^{x-1} - 1) C_2'\} (q v^{\frac{x-2}{2}})^{x-1}.$$

Hence, during the $x+1^{th}$ year, the whole possible loss of income will be:—

From expenditure of Proprietor, $a \{1 + l \Sigma_{y=x-1}^{y=1} (q v^{\frac{x+y-1}{2}})^{x-y}\}$.

From Capital removed or destroyed whose profits would have been expended without accumulation,

$$(m + m'q) C_2 - n (q - 1) C_2' + (n + n') A_2 + \lambda \{n (q - 1) C_2' - m C_2 - n A_2\}$$
$$- \lambda' (m' C_2 q + n' A_2).$$

From Capital removed or destroyed whose profits would have been accumulated,

$$\{\lambda [(m C_2 - n C_2') q v^{x-1} + n (C_2' + A_2)] + \lambda' (m' C_2 q + n_1 A_2) v^{x-1}\} (q v^{\frac{x-2}{2}})^{x-1}.$$

The greatest possible loss in this year, when v, l, m', n', λ', each $= 1$, and $\therefore m, n$ each $= 0$, will be

$$a . \frac{q^x - 1}{q - 1} + q^{x-1} (C_2 q + A_2);$$

and the least when l, λ, λ', each $= 0$,

$$a + (m + m_1 q) C_2 - n (q - 1) C_2' + (n + n') A_2.$$

As far as the destruction of Capital is concerned the investigation, of course, applies to the case of proprietors becoming absentees, and not of their continuing so.

The general effects of absenteeism may be thus enunciated. There will in all cases be a loss to the home-revenue in those direct taxes whose payment can be evaded by absence. There will, whenever there are duties on importation and not on exportation, be a further loss of the import-duty paid on the foreign productions which the proprietor consumed when at home.

Beyond this there will be a diminution of the aggregate income of the community whenever the capital that is disengaged by the absenteeism is forced into less profitable employments than those it previously occupied, and in no other case. When therefore the country from which the proprietor absents himself exports raw produce, there will generally, though not necessarily, be a loss beyond the loss to the revenue, and this loss will be accompanied by a general increase of rental. When it exports manufactures, there will not in general be any loss beyond that to the revenue.

There will however in all cases be this further and very important effect: though the income which the proprietor removes may be replaced, it must be replaced by labour, and there will therefore be substituted for the leisure class, which a part of that income maintained, a class who must by their own exertions produce the incomes on which they subsist; and there is nothing in the conditions of the problem to limit the extent to which the subdivision of income may, among the members of this class, be carried, or to fix the minimum that may be enjoyed by each.

It is necessary to the truth of these results, that the withdrawal of the proprietor should cause no removal of capital, that any part of the proprietor's income which was not expended should still be saved at home, and that no part should have been consumed without calling for the employment of capital.

In applying the result to any particular country, the first step is to decide how far, in the case of its absentees, these conditions are fulfilled; if they be not fulfilled, or if the individuals who remove had in any degree the qualities of productive labourers, the wealth of the community must be impaired by their absence; and the injury is capable of increasing with time to an indefinite extent.

J. TOZER.

CAIUS COLLEGE, CAMBRIDGE,
March 16, 1840.